The Thinker's Journal

THIS JOURNAL BELONGS TO

_____

The Thinker's Journal is a product of Mythical Legends Publishing
ISBN: **978-1-943958-91-7**
Copyright © 2018 by Mythical Legends Publishing
First Edition Published 2018

9 8 7 6 5 4 3 2 1

.

Lightning Source UK Ltd.
Milton Keynes UK
UKHW02f1129150818
327268UK00007B/41/P